IMAGES
of America

MARINES OF
WASHINGTON, D.C.

The November 10, 1954 dedication of the United States Marine Corps War Memorial drew a large crowd. The lone bugler playing taps puts the immense size of the monument in perspective, and the Washington Monument and Capitol Dome are visible on the Washington skyline.

IMAGES
of America

MARINES OF
WASHINGTON, D.C.

Mark Blumenthal

ARCADIA
PUBLISHING

Copyright © 2004 by Mark Blumenthal
ISBN 978-0-7385-1628-8

Published by Arcadia Publishing
Charleston, South Carolina

Printed in the United States of America

Library of Congress Catalog Card Number: 2004100202

For all general information contact Arcadia Publishing at:
Telephone 843-853-2070
Fax 843-853-0044
E-mail sales@arcadiapublishing.com
For customer service and orders:
Toll-Free 1-888-313-2665

Visit us on the Internet at www.arcadiapublishing.com

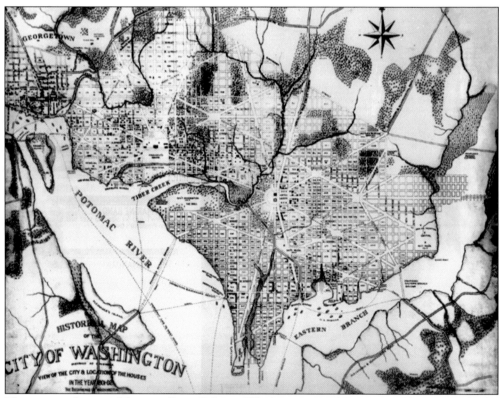

This map is a layout of the new Capitol City of Washington in 1801, prepared by Pierre Charles L'Enfant in 1791. Soon the United States Marine Corps would establish an enduring presence in the new Capital. (National Archives.)

CONTENTS

Acknowledgments   6

Introduction   7

1.   Early Years: 1801–1916   9

2.   World Wars: 1917–1945   37

3.   Post War: 1947–1970   57

4.   Modern Day: 1970–Present   93

Recommended Reading   128

ACKNOWLEDGMENTS

This book is dedicated to the Marines of our national Capital region, including those serving today as well as those who have served since the establishment of a permanent Marine presence in Washington in 1801. The pictures included in this history are representative of the incomparable service and dedication of the Washington Marines throughout the past 200 years. Therefore, this short work is by no means exhaustive. Those readers—whose interests are peaked—and those who desire to explore the varied history of the United States Marines in general—or more specifically, the Washington Marines—are encouraged to refer to the recommended reading or the resources of the Marine Corps Historical Foundation and the History and Museum branches of Headquarters U.S. Marine Corps. Lastly, a visit to the Marine Barracks at 8th and I or the Marine Corps War Memorial to view an evening or sunset parade is enthusiastically recommended and should be the highlight of any visit to the Capital region, which could be further complemented by a tour of the National Museum of the Marine Corps.

Many notes of appreciation are necessary for those who assisted in the completion of this work. Pat Mullen, of the Alfred M. Gray Research Center Archives; Lena Kaljot, of the Marine Corps Historical Center; MGySgts D. Michael Ressler and Dale R. Allen, of the U.S. Marine Band Library; SSgt Kristin S.D. Merger, U.S. Marine Band Public Affairs; Sgt Leah Cobble, Marine Barracks Washington, Public Affairs; and Maj. Matt Mowrey, HMX-1, were instrumental in the compiling this photographic history. MGySgt Ressler is worthy of special recognition for his previous work, concerning the history of the U.S. Marine Band, which was referred to extensively, and for sharing his extensive expertise in all things musical. Unless otherwise captioned, all photographs are official U.S. Marine Corps photographs.

A continuing note of gratitude is offered to the staff at Arcadia Publishing for their constant support in assisting me in sharing my interests in old photographs of Marines of the "Old Corps" to the present day. Thank you to Susan Beck for her always impeccable assistance as editor and to Katie White for her efforts as publisher.

As always, my wife, Anne Marie, deserves my continual heartfelt appreciation for supporting me beyond just this work. Also, her assistance in narrowing down the wide variety of photographs to those most relevant to the scope of the project and her insight into captioning were invaluable.

INTRODUCTION

The presence of the United States Marine Corps in Washington predates the establishment of Washington as our Nation's Capital, since Marines had established a small foothold at the Washington Navy Yard during the late 1790s. However, with the planned move of the Capital from Philadelphia to the newly formed District of Columbia, the Marine Corps began to look for a more permanent location in the Capital.

The last day of March 1801, a Sunday, Commandant of the Marine Corps, Lt. Col. William Ward Burrows was accompanied by President Thomas Jefferson to select a location for what would become Marine Barracks Washington. The President desired a location that would allow the Marines to come to the defense of the Capital and to be able to provide support to the Navy Yard. Riding to the southeast, the President selected a location within easy marching distance of the Capitol. Construction on the Marine Barracks began shortly thereafter, and once complete, the Washington Barracks became home to several companies of Marines and the U.S. Marine Band.

Washington Marines played a key role in the War of 1812 when they were hastily summoned to Maryland as part of the forces assigned to stop the British march on Washington. These forces could not stop the superior British forces, yet their delaying action allowed the evacuation of the Capital. A Marine legend declares that the British, out of respect for the spirited Marine defense at Blandensburg, Maryland, spared the home of the Commandant of the Marine Corps while the White House and unfinished Capitol were torched. Thus, the home of the Commandant lays claim to being the oldest continuously occupied government residence in the district.

In 1820 Archibald Henderson began his long tenure as the Commandant of the Marine Corps during which time he further cemented the position of the Marine Corps not only in the Capital, but as a force in readiness. The United States, being a country distrustful of standing armies, was disposed to task the small United States Army to garrison duty at frontier outposts thus leaving the Marines, by default, as the only "soldiers" in close proximity to the Capital. This would prove valuable to the Marine Corps' aspiration for permanence in the years to come.

Commandant Henderson, desiring to see the Marines as more than a force subordinate to the Navy, did much to ensure Washington politicos were inculcated as to the usefulness of a Marine Corps. Henderson's personal leadership was demonstrated numerous times in his many years as Commandant. In 1832, Henderson personally led a hastily formed battalion of Marines, drawn from the Navy Yard and Washington Barracks Marines, on a campaign against the Seminole Indians in Florida, and in 1857 Henderson led Marines in quelling riots in the city. For Henderson's skillful display of the utility of a Marine Corps and for his 39 years as Commandant, Archibald Henderson's is rightfully considered the "Grand Old Man of the Marine Corps."

On October 16, 1859, the town of Harpers Ferry, Virginia, awoke on a slumbering Sunday morning to find that abolitionists under the radical John Brown had attempted to seize the Federal Arsenal in order to foment a slave uprising, but instead, when thwarted by the local militia, had barricaded themselves in a firehouse. The local militia sent a request to Washington for Federal Troops. The Federal Government called once again on the U.S. Marines as Federal "troops in residence" and readiness, sending Marines from the Washington Barracks. Lt. Col. Robert E. Lee was also summoned to take control of all forces at Harpers Ferry. Upon his arrival, Lee found the militia balking at assaulting the building, pointing to federal jurisdiction. Lee therefore offered the honor of assaulting the building to the Marines, who stormed the building and captured Brown and his followers, with the loss of one private. In the American Civil War, Marines from the Washington Barracks participated in the Battle of Bull Run and acted as raiding and landing parties against Confederate forces who blockaded the Potomac River early in the war. However, the needs of the Navy for Marines to serve on ships' detachments limited the involvement of Marines in the Civil War in general, as well.

In the last 20 years of the 19th century the Marine Band, which was already well known in the Capital due to performances for numerous Presidents and wide ranging public performances, would see their popularity rise across the nation due to the leadership of John Phillip Sousa. Sousa, having apprenticed with the band, assumed duties as Director in 1880 and began a transformation, which included newer instrumentation and rigorous practice schedules. In an era of band stands across America, Sousa's authorship of over 100 marches, including "The Stars and Stripes Forever" and "The Washington Post," tours of the United States, and some of the first recordings of the Marine Band on Edison's new phonograph brought critical acclaim to the band already known as the "President's Own."

World War I saw a large expansion of the Marine Corps concentrated on the East Coast. Marine recruit training was shifted from the Washington Barracks to Parris Island, South Carolina. Officer and advanced infantry training were moved to the newly established Marine Base at Quantico, Virginia. The first instance of women in the Marine Corps was also driven by wartime expansion when the Marinettes were formed and began to fill administrative and support roles primarily with the Marine Headquarters. World War II drove an exponential increase to the Marine Corps, growing to five complete Marine Corps Divisions with supporting Marine Air Wings and logistical units, which likewise expanded the Headquarter's functions in Washington.

In 1950 the transformation of the single most recognizable image from World War II, the flag raising on the Island of Iwo Jima, captured by AP photographer Joe Rosenthal, began when Artist Felix de Weldon began work on the U.S. Marine Corps War Memorial. This bronze sculpture captured the essence of the photograph and is one of the most popular photo backdrops for D.C.

Military formations and parades have always been a means of assembling men for battle, issuing orders, and marking special occasions. As such, parades at the Washington Barracks have been normal occurrences. During the 1930s, the Barracks had begun to allow spectators to observe these parades informally, but during the late 1950s these parades began a transformation into performances combining aspects of three military rituals: the parade, the lowering of the colors or "retreat," and the pageant. Marine Barracks Cdr. Leonard F. Chapman, who would become the 24th Commandant, planned an "Evening Parade" open to the public at the Barracks, on Friday evenings in the summer months, which would become a unique and patriotic tradition of the "Oldest Post of the Corps."

Beginning in the 1960s, Presidential transportation began to look towards helicopters. This role is performed exclusively by Marine helicopters from HMX-1, the Presidential Support Squadron. Flying distinctively "Marine Green" SH-1 and UH-60 Sikorski helicopters with distinctive "white tops," these aircraft are now synonymous with arrivals and departures of the President of the United States on the south lawn of the White House and during Head of State visits. "Marine One" has become the most modern of Presidential images and one most closely associated with the Marines.

The modern Marine Corps presence in the Nation's Capital continues with a level of distinction synonymous with that of Marines who served the Capital region over the past 200 years. The duties upheld by these earlier Marines have become traditions that are still seen and heard today. Washington Marines served in Operation Desert Storm in similar fashion to Archibald Henderson's Indian Wars. The Esprit de Corps of the Washington Marines is available for all to witness during parades held at the Marine Corps War Memorial and Marine Barracks during which the stirring marches of John Phillip Sousa can be heard with instrumentation as he intended. And, as the sun sets in places around the globe where Marines stand ready to answer the Nation's call, and the Nation's colors are struck at sunset, the flag atop the 70-foot staff of the United States Marine Corps War Memorial is flown 24 hours a day in testimony to the valor of Marines who have gone before and those who continue to sacrifice today.

One

EARLY YEARS
1801–1916

President Thomas Jefferson, a political opponent but personal friend of Marine Commandant William Ward Burrows, participated in the selection of the site for the Washington Barracks. Jefferson desired a location that would allow the Marines to protect both the Capital and the Navy Yard and selected square 927 southeast. (Painting by Col. Charles Waterhouse, USMCR.)

Washington Marines participated in the defense of Washington during the War of 1812, shown here at the Battle of Blandensburg, Maryland. In recognition of their valor, according to Marine Corps lore, the British Commander spared the home of the Commandant of the Marine Corps, while the Capitol and White House were both burned. (Painting by Col. Charles Waterhouse, USMCR.)

Commandant Col. Archibald Henderson's 39 years service as Commandant earned him the moniker "Grand Old Man of the Marine Corps." Henderson believed in personally leading Marines and did so in the Seminole Indian Wars in 1836–1837 and, according to Marine legend, left a note tacked to his door stating, "Gone to Florida to fight the Indians. Will return when war is over." (Painting by Col. Charles Waterhouse, USMCR.)

Commandant Henderson's October 15, 1823 wedding was quite the social event, and this painting depicts the formal introduction of Henderson's bride, the former Anna Maria Cazerone of Alexandria, Virginia, to Washington society. (Painting by Col. Charles Waterhouse, USMCR.)

In January of 1857, a "gang of toughs" in Washington attempted to interfere with mayoral elections by roughing up voters and seizing a cannon from a local armory. Marine Commandant Henderson, armed only with a top hat and cane, faced down the cannon as it was being directed on the group of Marines sent to halt the disturbance. (*Frank Leslie's Illustrated Newspaper.*)

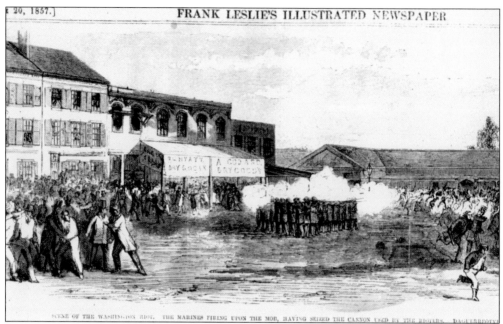

SCENE OF THE WASHINGTON RIOT. THE MARINES FIRING UPON THE MOB, HAVING SEIZED THE CANNON USED BY THE RIOTERS. DAGUERREOTYPE

Marines of the Washington Barracks are forced to fire on the crowd, as local police were unable to cease the rioting. (*Frank Leslie's Illustrated Newspaper.*)

In 1859, abolitionist John Brown attempted to seize the Federal Arsenal at Harpers Ferry, Virginia. Marines from Washington were dispatched and placed under the command of Lt. Col. Robert E. Lee. The Marines successfully stormed the firehouse where Brown and his followers were barricaded. Brown was later hanged, becoming a martyr to the abolitionist movement.

This lithograph shows Marines returning to the Marine Barracks from the Capitol. On the eve of civil war, military activity in the Capitol was increasingly evident and Marines stood ready to protect federal interests and property. (*Harper's Weekly.*)

This is the earliest known photograph of the United States Marine Band, posed in 1864 on the parade deck in front of the Commandant's home. The band, formed in 1798 in conjunction with the establishment of the United States Marine Corps, is America's oldest professional musical organization.

The Marines of the Washington Barracks are pictured here, in formation in front of the old Center House, in approximately 1860.

These Washington Navy Yard Marines wear the uniform typical of the Civil War era. Note the lengthy bayonets of the day and the Officers' sword.

An 1864 formation of Marines stand in front of the Commandant's home.

Shown approximately 1866, this young apprentice is a fifer with the U.S. Marine Band. It was not uncommon for young boys to apprentice with the band, typically as fifers and drummers.

John Phillip Sousa began his long career with the band as an apprentice. Sousa desired to join the circus but his father instead persuaded him to join the Marine Band as an apprentice at the age of 13.

John Phillip Sousa, having assumed leadership of the band in 1880, poses in front of his Marine Band on August 23, 1882, in Cape May, New Jersey, during the annual outing of the Washington Light Infantry. This photo is the earliest known of Sousa with the Marine Band.

An interesting photo of the late 1880s shows off Marine Band uniforms, which included spiked helmets. Several admirers of the band can be seen including themselves in the photo through the windows and doorway.

The company of the USS *Sussex*, assembled on deck in 1888, is typical of Marine shipboard detachments at the beginning of the steamship Navy. The drummer stands at the ready to drum commands to the assembly.

The Marines pictured here are on the grounds of the U.S. Naval Academy.

Pictured is a Marine formation on the parade deck of the Washington Navy Yard, c. 1889. The spiked helmets, popular at the end of the Prussian War, have given way to uniforms more closely resembling those of the Civil War era.

18

This Marine Colonel in plumed spiked helmet could easily be confused with a German or Prussian officer of the era, save the distinctive Marine Corps emblem, consisting of an eagle, globe, and anchor on his helmet.

Maj. Gen. Charles Heywood, center, the ninth Commandant of the Marine Corps, is shown below in late 1891, with his staff. Heywood distinguished himself during the Civil War and in the seizure of Panama City from rebels in 1885. Heywood was the first Major General Commandant.

By the time this 1890 photo of John Phillip Sousa was taken, he was the undisputable "March King," having composed over 40 marches such as "Semper Fidelis," "The Thunderer," and "Washington Post". In 1891, Sousa would begin annual concert tours of the United States that continue today.

This is the cover of the concert program used for the first Marine Band concert tour, in 1891. The fanciful cover, in full color, shows both the White House and the Capitol, locations synonymous with the Marine Band. The emblem at lower right is the original Marine Corps emblem, which still adorns the buttons of Marines' uniforms.

The 1892 Band pose in blue dress uniforms. Of particular interest is the variety of beards and mustaches popular during the time.

A poster announces the Southern tour of the Marine Band under the directorship of Franco Fanciulli, who led the Band from 1892 until 1897. The band was commonly referred to as the Nation's Band, due to the popular nationwide tour schedule.

These Washington Navy Yard Marines are on review in the "present arms" position in this 1896 photograph.

This 1896 photograph of barracks officers includes, from left to right, Capt. P.S. Murphy, Capt. T.N. Wood, 2d Lt. J.H. Russell, 2d Lt. L.J. Magill, 2d Lt. J.T. Meyers, and 2d Lt. J.H. Pendleton. Russell would become the 16th Commandant in 1934.

The original sheet music cover of John Phillip Sousa's most famous march, "The Stars and Stripes Forever." Written in 1896, this march was declared the "March of the United States," by the Proclamation of President Ronald Reagan on December 11, 1987.

The Marine Band marches in the 1897 Inaugural Parade of President William McKinley. The Marine officer at the lower right, distinguished by the Mameluke sword he wears, bears a resemblance to John Phillip Sousa; however, Sousa left the Marine Band in 1892 to form his own band.

The Marine Band performs in front of the U.S. Capitol in 1899. These concerts, quite popular with large crowds, continue today.

Members of the Marine Barracks and the Marine Band pose in front of the original band hall. Construction on a new band hall, with room for audiences to attend concerts, began in 1905.

The Capitol concert continued to gain in popularity and crowds began to grow because of the availability of automobiles, such as the early Ford Model-Ts visible.

In this photograph, the Marine Band is pictured at the Capitol. Dir. William H. Santlemann, center, director from 1898 to 1927, created an orchestra within the band by requiring musicians to be proficient on both a string and wind instrument.

This rest stop on the 1901 tour provided an opportunity for this impromptu photograph. The Marine on top of the coach wanted to ensure he would stand out in the picture.

The Marine Band makes a formal appearance in the 1904 World's Fair parade. Held in St. Louis, Missouri, at the peak of the band era, competition was high because former Marine Band Dir. Sousa was present, leading his own band, and many military bands participated.

The Marine Band assembled for a rehearsal in an exhibition hall during the 1904 World's Fair.

showing Old Center House.

This photograph shows the Marine Detachment in 1905. "Old Center House," left, would soon be demolished, much to the disappointment of barracks officers, during a major construction project.

The first brick is laid for five officers quarters, which will line 8th Street.

The remains of the old arcade and the band hall are visible. Both will soon be demolished.

The building on the far left nearing completion is the new "Center House," which will serve as an officers quarters and mess.

Finishing touches and clean-up on the new officers quarters is nearly complete in this 1908 picture.

The old band hall has been demolished, and the lower level of the new hall nears completion in this 1906 view.

In the rear of this photo is the arched stage of the new band hall.

A Marine guard stands his post in front of the band hall arcade's arches.

A worker stands on the peak of the nearly completed band hall, with the western Washington skyline in the background.

The completed band hall arcade is shown with work begun on the construction of a gate into the parade deck.

A view of the completed officers houses from the corner of 8th and I Street, c. 1910, also shows the streetcar tracks along 8th Street.

Pictured are the officers of the School of Application. This school, formed by Commandant Heywood, was relocated to Quantico and is the forerunner of the Basic School, located at Marine Corps Base Quantico today.

The crew of the USS *Stratford* stands in formation at the Washington Navy Yard. The First Sergeant holds an 1859 non-commissioned officer's sword. Only Marine NCOs are authorized to carry swords when commanding Marines.

This 1911 photograph, used on fliers announcing band appearances, is interesting in that the band was superimposed on top of a photograph of the White House.

Marines are on the firing line at Camp Winthrop, Maryland. All Washington Marines used this range until the establishment of ranges to the south, at Quantico.

This Marine takes aim with a Krag-Jorgenson rifle by wrapping a hasty sling around his arm. The boondocker-style boots he wears survive in form and are still authorized for field grade officers and staff non-commissioned officers.

Marines are shown pulling butts. When the crack of the rifle projectile is heard overhead, the Marine will run the carriage down, "spot the round" with a marker showing the impact location, and raise the target back up for the next shot. The scoring value of the shot is relayed to the shooter using the "disk" seen to the left.

These Marines are scoring the match by observing the spotting calls as they are "disked" from the butts.

The trombone section of the Marine Band pose for a seated portrait. The uniforms worn by the Marine Band have changed only slightly from those worn in this 1915 photograph.

Two

WORLD WARS
1917–1945

This photograph illustrates the 5th Regiment on Parade in Washington in 1917. With American involvement in World War I looming, a significant expansion of the Marine Corps was occurring both in Washington and the newly established Marine Barracks in Quantico, Virginia, 30 miles to the south.

This view of the Commandant's home, c. 1917, shows the graciousness of its Georgian-Federalist design.

The Barracks' Parade Deck is pictured here, as viewed from the Commandant's residence. The completed band hall is seen to the right of the photograph.

World War I brought the first women into the ranks of the Marine Corps. The Marinettes performed mostly administrative duties, primarily in the Washington area, and are shown here hanging recruiting posters.

These recruiting posters are visible in the photograph at the top of the page. Left is the German name for Marines, "Devil Dogs!" Right is a now famous poster by James Montgomery Flagg.

Assistant Secretary of the Navy Franklin Delano Roosevelt and Commandant of the Marine Corps George Barnett share watermelon with staffers in Washington, c. 1917. Roosevelt and Barnett were key figures in the establishment of a permanent Marine Base at Quantico.

Illustrated here, the Fifth Regiment prepares to march in a victory parade in Washington in 1919.

The Fifth Regiment marches in review down Pennsylvania Avenue.

The Fifth Regiment battle colors are presented in front of the White House.

The 1920 Marine Band is standing on the steps of the Thomas Jefferson Building of the Library of Congress. Thomas Jefferson is credited with bestowing the Marine Band with the title "The President's Own."

This photograph offers a rare view of an indoor concert in the Marine Band Hall in 1920. Concerts, such as this one by the Marine Orchestra, were very popular and seating in the hall was extremely limited.

This Marine of the Quarter Master Corps poses proudly beside his Mack truck in the 1920s. The Quarter Master Corps was housed on the grounds of the Washington Navy Yard.

A G Street view depicts the front of the Commandant's quarters in the mid-1920s. The quarters' front faces the north, and the southern exposure looks onto the parade deck.

Two sentries man the 8th Street gate of the Marine Barracks.

This view of the barracks, in approximately 1920, shows a variety of period vehicles. The building to the left is the Center House, although it is no longer actually in the center of the parade deck.

This photo offers a view of the muddy parade ground in the 1920s showing the arched walkway of the arcade and the rampart centered on the parade deck directly behind the flagpole.

Members of the Marine Band stand at the railway station in South Bridge, Massachusetts, during the 1921 tour. The band often traveled by rail on the annual tours. The Marine on the left obviously favors his right side in this 1920 photograph.

Bill Jr., son of Old Bill, announces an appearance of the Marine Band in Newark, New Jersey, during the 1921 tour.

Soloists of the Marine Band pose for a formal portrait in approximately 1920. Soloists were recognized for their musicianship and great popularity with Washington, D.C. audiences.

Band tours were performed on a sponsorship basis, as no funding was available. Typically, Rotary Clubs, chambers of commerce, and businesses provided funding and transportation, such as a lift in this 1920s touring car.

This picture shows a July 16, 1921 concert on the south lawn of the White House.

The Marines participated in many Civil War reenactments during the 1920s. Marines from Quantico and Washington are shown here, in 1924, on the march to the Gettysburg, Pennsylvania battlefield.

In 1924, the White House, under President Woodrow Wilson, was temporarily re-located to the Gettysburg battlefield, in the large white tent, to commemorate the 60th anniversary of this historic battle.

A 1925 Capitol Concert is about to begin under the directorship of William H. Santlemann.

Maj. Gen. Smedley Butler poses with a trio of bulldogs at a 1928 Quantico Marines football game in Washington. Butler was largely responsible for the English bulldog being adopted by the Marines. The bulldog in the center is the original "Mr. Jiggs."

Smedley Butler, shown leading a cheer for the Marines, was responsible for gaining exposure for the Marine Corps through football competitions featuring the Quantico Marines. The Quantico Marines posted an impressive record against college teams.

This photograph is of the 1927 "surprise serenade." The Marine Band has a long tradition of surprising the Commandant with a concert New Year's morning. Shown here are Maj. Gen. John Archer Lejeune and Dir. William H. Santlemann, during Santlemann's last serenade with the band.

General Lejuene's emphasis on the professional development of the Marines continues to have lasting impact on the Marine Corps. He is shown here in 1927; the wreath around the Marine Corps emblem on his cover distinguishes him as the Commandant of the Marine Corps.

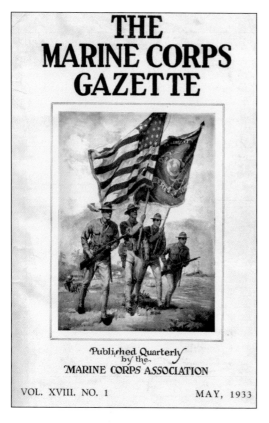

THE MARINE CORPS GAZETTE

Published Quarterly by the MARINE CORPS ASSOCIATION

VOL. XVIII. NO. 1   MAY, 1933

The *Marine Corps Gazette* is a professional publication founded by Marine Officers in Quantico, Virginia, in 1917. Publishing of the *Gazette* later moved to the Washington Navy Yard and returned to Quantico as part of the Marine Corps Association in 1974.

Marines, in blue white dress, march in a 1930 Washington parade, executing "eyes right," and have just noticed the spectators who have inadvertently entered their line of march.

The Washington Barracks Marines, shown here, are in formation in approximately 1930.

This photograph was taken at the New Year's 1942 surprise serenade for Commandant Clifton Cates. The issuance of "grog," an old naval service custom, survives only at these events; the Commandant serves hot rum toddies to the Bandsmen after the serenade.

President Roosevelt is pictured at the April 13, 1943 dedication of the Jefferson Memorial. The tidal basin is behind the President as he faces the memorial. A number of broadcasting companies are airing the Marine Band's performance and the President's remarks on live radio.

This photo provides an interesting angle on a Marine Band performance in 1944. The Director on the platform is William F. Santlemann, who joined the band in 1923 while under the directorship of his father, William H. Santlemann, and who led the band until 1955.

The Marine Band is marching on the grounds of the U.S. Capitol in 1944 in this photograph.

The Marine Band concert shown in this photo took place at the east front of the U.S. Capitol in 1944. Audience members are being seated on the Capitol steps.

Three

POST WAR
1947–1970

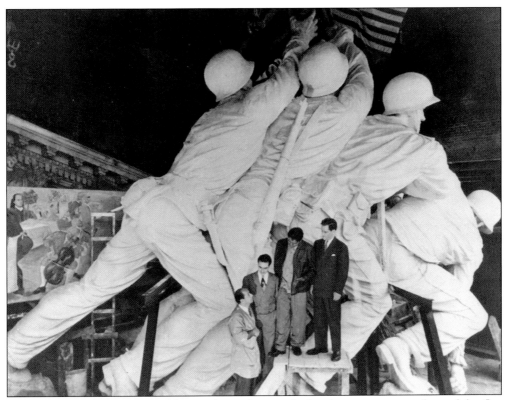

The sculptor, Felix W. de Weldon, is shown here with the three serving members of the flag raising on Iwo Jima. Pictured in front of the plaster model of the Marine Memorial, from left to right, are Felix de Weldon, Pfc. Rene Gagnon, Pfc. Ira Hayes, and PhM 2/c John H. Bradley.

De Weldon is shown looking at the face of one of the flag raisers. The plaster model would soon be carefully cut into sections, to be cast in bronze.

Two Sergeants marvel at the size of this cast boot. The completed castings, weighing some 20 tons, were shipped by truck convoy from the foundry in the Bronx, New York, for assembly in Arlington, Virginia.

The completed casting of Cpl. Harlon H. Block planting the base of the flagstaff has been placed into position. The flagstaff, when completed, was over 70 feet tall, and this piece alone weighs over 5 tons.

Three cranes were required to lift pieces, which were then welded together to complete the memorial. The rough Swedish granite base is 6 feet tall and is inscribed with the names of battles in which Marines have fought since their establishment in 1775.

An M-1 rifle is lifted into place. An actual M-1 rifle is 42 inches long, in comparison to this cast rifle of 16 feet.

Sculptor Felix de Weldon, left, looks on as an M-1 carbine is hoisted.

President Dwight D. Eisenhower arrives at the dedication of the Marine Corps War Memorial on November 10, 1954, the 179th anniversary of the founding of the U. S. Marines.

From left to right are President Eisenhower, Vice President Richard M. Nixon, and Gen. Lemuel Cornick Shepard, the 20th Commandant of the Marine Corps.

The crowd gathers for the dedication. Located just across the Memorial Bridge in Arlington, Virginia, the memorial cost approximately $850,000 and was funded solely by Marines and private donations.

With the end of World War II, Marine Barracks again became the focal point of the Marines in Washington. Many foreign military services and political dignitaries visited, such as this 1947 review by the Commandant of the British Royal Marines.

At the end of World War II, women were, for the first time, made a part of the regular Marine Corps and were finally issued uniforms specifically designed for women. From left to right are the dress blue, summer, and service uniforms.

Marine Barracks, Arlington, Virginia, was first established as a barracks for women Marines during World War II. The post was renamed Henderson Hall in honor of Archibald Henderson and supports Headquarters Marine Corps activities. The edge of Arlington National Cemetery is visible at the lower left.

This 1949 aerial places the Marine Barracks quadrangle in perspective to the surrounding southeast Washington neighborhood. The diagonal avenue to the upper right is Pennsylvania Avenue South East and Virginia Avenue South East appears at the lower left.

A tighter shot of the barracks shows the quadrangle, formed with the Commandant's home on the north, the band hall on the south, and the officers quarters and barracks offices on the west and east sides of the quadrangle respectively. Center House is the house closest to the band hall.

Members of the Marine Band prepare for departure to perform at the 175th anniversary of the historic Revolutionary War battle of Lexington and Concord.

The cornets, trumpets, and percussion of the Marine Band perform in the Band Hall , c. 1950.

Wartime expansion necessitated construction of a Navy Annex. Elements of Headquarters U.S. Marine Corps are housed here, in the Pentagon, and at Quantico Marine Corps Base.

Parades have always been a part of martial service, especially at the Marine Barracks. This 1950s parade is the forerunner of what will soon become the "Evening Parade" at the barracks and the Marine Corps War Memorial.

Radio broadcasts of Marine Band performances began in 1922 and continued throughout the radio era into the early 1960s.

A unique view of a New Year's serenade. Businesses on 8th Street are visible to the west of the Commandant's home over the palisade.

The Fort Henry Guards of Ontario, Canada, shown here in 1955, were frequent visitors to the Marine Barracks, beginning in 1954. The Guard, a 19th-century military tactics demonstration unit, performs in concert with the Marines at the barracks, and the Marines reciprocate by appearing at Fort Henry.

The Fort Henry Guards advance their firing line, demonstrating British tactics of the 1860s.

The Guards have just fired a volley and begin to advance. In addition to their squad tactics demonstration, the guards also perform a sunset ceremony similar to that of the Marines.

The Fort Henry Guards' ceremonial unit performs with the Marines at the Iwo Jima Memorial.

The band prepares for another busy touring season by packing uniforms. Marine Band members are recognizable by the musician's lyre on their rank chevrons; Marines in general have crossed rifles on their rank insignia. The Marine Band is composed of professional musicians recruited specifically for the band.

Each year the band selects different localities throughout the country to appear, in order to provide multiple opportunities to the American public to witness a performance. This photograph is from the Holland, Michigan tour at the annual Tulip Festival.

A state border—in this case Nevada, in 1951—is always a good opportunity for the band to assemble for a photograph while taking a break from traveling.

Marines place a wreath at the Tomb of the Unknown Soldier each year on Memorial Day.

Members of the Black Watch, a Scottish Regiment of the British Army, pay a visit to the barracks in 1959. Shown here with their respective Drum Majors are a Black Watch soldier and a Marine Public Affairs representative, presenting a Marine snare drum.

This unique angle focuses on Band Director William F. Santlemann.

This publicity photo, c. 1950, of the Marine Band was used in press releases for concerts on the Capitol lawn.

The Marine Corps Institute (MCI), founded as a vocational school by order of Major General Commandant Lejeune at Quantico in 1920, was later moved to the 7th and G Street location, shown here. MCI provides Marines for the ceremonial detachments at Marine Barracks.

Marines depart the Marine Corps Institute for the barracks in 1950. MCI Marines participate as ceremonial marchers for Evening Parades and serve other functions at the Marine Barracks.

Eisenhower is sworn in as President of the United States of America. The Marine Band is seated just below the inaugural platform. The band will strike up "Hail to the Chief" when the oath of office is complete and will later participate in the Inaugural Parade.

Here the Marine Band poses on the Capitol steps. This view is similar to that on the cover photograph. The band often poses at the Capitol for publicity photographs.

This facility is home to MCI logistics at the Washington Naval Ship Yard. MCI provides hundreds of correspondence courses in both professional and vocational topics and has served over five million Marines since 1920.

The Marine Corps Institute later moved to this location at the Washington Navy Yard.

The first use of helicopters, like this executive support helicopter (Sikorski SH-3H) above the Capitol to support the President of the United States, occurred in 1957: a Marine UH-34 was used to ferry President Eisenhower to Air Force One when his presence was immediately required in Washington, D.C.

The Marine Corps Drum and Bugle Corps are shown marching in the Inaugural Parade for President John F. Kennedy.

President Kennedy departs from "Marine One" in 1961. All Marine Helicopters transporting the President are designated Marine One. The helicopter here is a Sikorsky VH-3D, Sea King.

Caroline Kennedy sits on First Lady Jacqueline Kennedy's lap on board Marine One.

Defense Secretary Robert McNamara speaks as President Kennedy looks on. A military honor guard is formed in the background, as are several Marine UH-34 helicopters supporting the visit.

The Marine Band stands in front of the south portico on the south lawn of the White House in 1962. The Marine Band participates each year in the annual Easter Egg Roll.

The Ceremonial Marchers of Marine Barracks 8th and I are in formation at the Marine Corps War Memorial.

Here is an artist's rendering of the proposed Marine Corps Museum, which was to be located adjacent to the Marine Memorial. This plan, however, was never realized.

Formed in 1948, the United States Marine Corps Silent Drill Team performs an intricate routine, based upon regulation drill movement and the "manual of arms," totally without verbal commands. Instead, they rely solely on the discipline instilled by hours of daily practice.

A Marine Corporal performs a distinctive rifle inspection, which when completed, will include an awe inspiring, seemingly effortless, spinning rifle, flipped behind the back and over the shoulder of the inspector, and snatched, out of mid-air, by the Marine being inspected.

The Marine Band's versatility is apparent in the wide variety of band combos. Shown here is an accordion, bass, and violin trio.

A jazz combo, comprised of members of the Marine Band, plays a gig in 1963. The Marine Band is no longer able to perform such combos because of their many official duties and concert performance commitments.

The Marine Band leads the way in the 1964 Inaugural Parade for President Johnson.

Lady Bird Johnson precedes President Johnson, deplaning from Marine One. Johnson was reportedly quite fond of the "presidential" image and power projected by Marine One flights.

President Johnson and a foreign visitor are pictured on board Marine One in 1965. Flights with foreign dignitaries aboard Marine One are often a key highlight of visits to the Capitol and allow the President to point out the multitude of Washington's landmarks and monuments.

Pictured in this photograph is the Commandant of the Marine Corps Gen. Wallace M. Greene Jr. with the winners of the 1967 Super Squad competition. This competition tests rifle squads from each of the Marine Divisions on a variety of military skills. The winning squad is considered the best rifle squad in the Marine Corps.

A Marine Corporal poses with Corporal Chesty, the Marine Corps mascot. Chesty makes an appearance at each Evening Parade, which is especially delightful to younger members of the audience.

Sergeant Major Vousa of the Royal Constabulary, on the island of Guadalcanal during World War II, receives Honors at the Marine Barracks in 1968. Vousa was captured by the Japanese, but he escaped and helped to lead many patrols during the Guadalcanal Campaign.

The Marine Barracks is often the site of awards ceremonies. Here, the parents of Pfc. Douglas E. Dickie are presented with his posthumous Congressional Medal of Honor. Private First Class Dickie was awarded the Medal of Honor for actions in Southeast Asia.

The Fort Henry Mascot, David the Goat, is welcomed to the evening ceremonies by the Marine Barracks Adjutant. Goats are the mascot of many British and Canadian regiments.

Gun smoke wafts away as the Fort Henry tactics demonstration squad reloads after firing a volley during a performance at the Marine Memorial in 1969.

Above, Chesty is introduced to soldiers of the Fort Henry Guard during a tour of Marine Barracks Washington.

The Marine Barracks Color Sergeant points to a plaque commemorating the founding of the barracks in 1801.

The Drum Major of the Marine Band strikes a formal pose with Chesty, the Marine Corps Mascot. In addition to the red dress uniform of the Marine Band, the Drum Major wears a headpiece made of bearskin and a baldric across his chest, which is embroidered with the Marine Band crest. The Mace he carries is embossed with the names of Marine battles and campaigns and is used to signal commands to the marching musicians. Chesty, looking squared away in his Marine dress blues and knowing he is the center of attention of any Marine Barracks function, is unimpressed with the Drum Major's additional ornamentation.

Four

MODERN DAY
1970–PRESENT

The Drum Major leads the band in a "Sound Off" on the parade deck.

The Color Guard leads the formation while "Passing in Review" during a parade at the barracks. The official Marine Corps Battle Colors have over 50 battle streamers.

Lt. Col. Charles P. Erwin, Assistant Band Director, directs the Marine Band, with the assistance of two Washington area children, during a 1980 children's concert in the John Phillip Sousa Band Hall.

This picture shows the Marine Band trombone section during a 1970 performance.

President Richard M. Nixon and First Lady Patricia Nixon pose with Marine Band Director Albert F. Schoepper, prior to a state dinner for the Prime Minister of Great Britain, on December 17, 1970.

The Marine Band is often called upon to provide accompaniment for musical performances at the White House, such as this 1972 performance with Frank Sinatra.

This photograph illustrates the "March on of Troops" for an Evening Parade at the barracks in 1972.

The Marines assemble at "Present Arms," while the national ensign is lowered.

After the colors are secured, the Marine Corps Color Guard will present the Battle Colors to the spectators. The Marine ceremonial marchers are armed with World War II–era M-1 Garand rifles.

The highlight of any Evening Parade is the performance by the Marine Corps Silent Drill Team, shown here, performing their intricate and precise routine.

The Marine Corps Drum and Bugle Corps performs "Sound Off" during a parade at the barracks.

The Marine Sergeant, having completed his inspection of the M-1, exchanges this rifle for the rifle of the next Marine in the rank. Although the M-1 rifle weighs 10 pounds, these Marines make the complicated spins and tosses seem effortless.

This photographs shows the Pass in Review, as seen from above the Band Hall.

Band Dir. Jack T. Kline, Director of the Marine Band from 1974 to 1979, is pictured here in the recently dedicated John Phillip Sousa Band Hall. The dedication on October 7, 1974, was officiated by the Honorable J. William Middendorf II, Secretary of the Navy, and 25th Commandant Gen. Robert E. Cushman, Jr.

President Jimmy Carter and a foreign guest listen to a performance by a harpist of the Marine Band. The diversity of instrumentation within the Marine Band places them on par with virtually any major concert orchestra in the United States.

The Marine Corps Historical Center, located at the Washington Navy Yard, houses the History and Museum Division of Headquarters Marine Corps.

The Marine Historical Center also houses a number of historical displays of the Museum Division, such as this display of machine guns and the Marine Corps art collection.

The Marine Band pose on the steps and balcony of the south portico. Band Dir. John R. Bourgeois is center.

"The President's Own"
United States Marine Band
Colonel John R. Bourgeois, Director

WEST FRONT OF THE
U.S. CAPITOL BUILDING

The Marine Band is shown here in a formal pose.

This photograph was taken at President Reagan's 1981 inauguration in front of the Capitol. A joint services color guard, representing all the Armed Forces of the United States and the Marine Band, are formed below the President.

Marine One lands on the south lawn of the White House.

Marine Helicopter One facilitates Presidential trips throughout the United States and abroad, including this trip to New York City in 1983. The twin rotor aircraft is a CH-46 Sea Knight, which is used to carry support staff and members of the press covering the visit.

President Ronald Reagan and First Lady Nancy depart the White House on Marine One. The Marine standing by to secure the aircraft hatch is the crew chief of the aircraft.

President Reagan, center, acts as Reviewing Officer at an Evening Parade, on the occasion of the Change of Commandants. Left is Gen. Robert H. Barrow, the 27th Commandant of the Marine Corps, relinquishing command to Gen. Paul X. Kelley, the 28th Commandant of the United States Marine Corps.

The Evening Parade Pass and Review is shown from the Guest of Honor's perspective on the barracks' center walk. President Reagan personally requested a change of dates for General Barrow's retirement so that he would be able to preside.

President Ronald and First Lady Nancy Reagan arrive to host a White House reception.

The "Commandant's Own," the United States Marine Corps Drum and Bugle Corps, perform a "play off" for Commandant of the Marine Corps Gen. P.X. Kelley. The traditional play off occurs when an officer of the barracks departs for a new duty station or retires.

Gen. Alfred M. Gray, the 29th Commandant of the Marine Corps, speaks to a group of Marines. The official portrait of General Gray, rendered in the camouflage utility uniform—the only Commandant's that is so, is an interesting contrast to this photo of him wearing the dress blue uniform. Note the additional wreathing on his cover, denoting his position as Commandant, which is similar to that dating back to the 1860s.

This gathering of Marine General Officers is the General Officer Symposium, which is held annually and allows the senior leadership to address issues, identified by the Commandant, to further improve upon the success of the Marine Corps.

This photograph displays the Commandant of the Marine Corps' quarters around 1985. Over the years, the house has received extensive renovations, including the addition of wings and the conversion of the attic into living spaces.

The entry way of the Commandant's home is fittingly under the gaze of fifth Commandant Archibald Henderson, who occupied the home for 39 years, from 1820 until 1859.

The living room fireplace is flanked on the left by Alexander A. Vandegrift, 18th Commandant, and on the right by Thomas Holcomb, 17th Commandant. Formal portraits of past Commandants of the Marine Corps are visible throughout the home.

Archibald Henderson also presides at the head of the dining room table.

This view shows the fireplace and piano in the Music Room.

Buglers high above the arcade rampart sound the "fanfare," which will denote the beginning of a summer Evening Parade at Marine Barracks Washington.

The Marine Barracks Color Guard is entrusted with presenting the Marine Corps Battle Colors.

The Marine Corps Silent Drill Team marches into the spotlights for a demonstration of precision drill, executed without audible commands, demonstrating perfect timing.

Members of the Silent Drill Team are armed with M-1 Rifles and perform with fixed bayonets, demonstrating the precision and team work that has made them famous worldwide.

Two Marines of the Drill Team at port arms demonstrate perfect alignment and synchronization, while performing the "manual of arms" on the march.

The Ceremonial Marchers pass in front of the Reviewing Officer and execute "eyes right."

The Color Guard marches past, during the "Pass in Review," carrying the official Battle Colors of the United States Marine Corps. The Marine Corps Battle Colors are entrusted to the care of the Color Sergeant of the Marine Corps. The Color Sergeant carries the national colors when marching in parades and reviews.

Once again the professional depth and versatility of the Marine Band is apparent in the wide variety of musical styles offered, such as this 1986 Dixieland Band.

The Marine Corps Drum and Bugle Corps performs on the parade deck. The Drum and Bugle Corps consists of specifically recruited musicians and Marines recruited from field bands throughout the Corps. Unlike the Marine Band, Drum and Bugle Corps Marines undergo basic military training.

The fact that the Marines of Marine Barracks 8th and I are combat-ready has been evident since the founding of the barracks. These Marines prepare to deploy by saying goodbye to loved ones during Operation Desert Storm in 1991.

In the early 1990s, HMX-1 added the Sikorski VH-60 to their inventory in support of the Executive Flight mission. The VH-60 is more readily transported by cargo aircraft and is ideal for missions across the country or around the world. The venerable SH-3 shown is a modernized VH-3D.

Four saxophonists—from left to right, two alto saxophones, a tenor saxophone, and a bass saxophone—pose under the watchful eye of John Phillip Sousa, on stage in the Sousa Band Hall.

The President's Own performed 19 concerts in the former Soviet Union during 1990, including this performance in Leningrad. This was the first such performance in the Soviet Union.

The President's Own poses with President George H.W. Bush and First Lady Barbara Bush during a White House Christmas Reception in 1991.

Although there are no longer any Marines serving the Washington Navy Yard onboard ships' detachments, Marines from the Washington Barracks keep the association alive by participating in ceremonies along side the U.S. Navy in the Navy Yard.

Gen. Charles C. Krulak, the 31st Commandant of the Marine Corps, presents the command to the Guest of Honor at a 1996 ceremony at Marine Barracks.

July 11, 1998, marked the occasion of the 200th anniversary of the United States Marine Band. The President's Own celebrated this noteworthy occasion in a concert and ceremony on the south lawn of the White House.

"The President's Own"
United States Marine Band
Colonel John R. Bourgeois, Director

THE U.S. MARINE BAND ON STAGE
IN THE JOHN PHILIP SOUSA BAND HALL
HOME OF "THE PRESIDENT'S OWN"

The full Marine Band, including orchestra elements, is shown in this photograph.

The Bicentennial Celebration of the founding of Marine Barracks 8th and I was marked by special displays and tours of the Barracks grounds, including Center House and the Commandant's home and gardens.

The Marine Corps celebrates the anniversary of its founding on November 10, 1775, each year with birthday balls and ceremonies on posts around the world. Here, a uniform pageant is held in conjunction with the Marine birthday, and historical uniforms from the founding of the Marine Corps are displayed.

Marines of the Washington Barracks participate in internments at Arlington National Cemetery. Here the Marine Corps Drum and Bugle Corps and an Honor Guard lead a "full military honors" procession.

Marine body bearers remove the flag draped casket from the caisson.

The flag is removed from the casket and will be folded properly into a tight tri-square before being presented to family members with the thanks of "a grateful nation" for the deceased's service to our country.

A composite bugle and trumpet section plays taps.

A seven-man firing detail prepares to fire three volleys for the traditional 21 gun salute. This tradition dates back to 1818 in the Naval Service, when there were 21 states, and has continued into the present as three rifle volleys of seven weapons.

The Marine Corps Battle Color Detachment, consisting of the Marine Corps Drum and Bugle Corps, the Silent Drill Team, and the Battle Colors, participated in the prestigious "Military Tattoo" at Edinburgh Castle, Scotland, during the 2002 season. Here, the Silent Drill Team begins their performance. (Photograph by Sgt. Leah Cobble.)

The Drum and Bugle Corps perform during the Tattoo for Queen Elizabeth II, who attended the performance as part of "Her Majesty the Queen's Golden Jubilee." Over 300,000 spectators attend the three-week tattoo season, and over 100 million television viewers enjoy the annual military pageant on television. (Photograph by Sgt. Leah Cobble.)

"THE PRESIDENT'S OWN"
UNITED STATES MARINE BAND
LtCol Timothy W. Foley, Director

In John Philip Sousa Band Hall
Home of "The President's Own"

In 1996, Col. Timothy W. Foley was selected to lead the band as the 26th Director.

"THE PRESIDENT'S OWN"
UNITED STATES MARINE BAND

In Concert

Colonel Timothy W. Foley, Director

Colonel Foley takes the podium as the Director of the Marine Band.

The Marine Corps Heritage Center—
Promoting the History of Our Corps

"*Leatherneck*, the Magazine of the Marines," was formed by enlisted Marines at Quantico during World War I. The magazine was such a success that the Marine Corps made it official and moved publishing to the Marine Barracks in Washington. The magazine moved back to Quantico in the 1970s. Publishing of both *Leatherneck* and *Marine Gazette* are now part of the Marine Corps Association. This 2003 cover announces the groundbreaking for the National Museum of the Marine Corps. (Chadbourn and Associates.)

This is an architect's rendering of the National Museum of the Marine Corps. The museum depicts the history of the United States Marines through a variety of interactive and multimedia displays. The completion of this museum realizes the efforts of the Marine Corps Historical Foundation and its many Marine Corps supporters to provide a national museum to share the Marine history with the American public. (Fentress Bradburn Architects, LTD.)

This contemporary view of the parade deck of Marine Barracks 8th and I shows the viewing stands for spectators of the Evening Parades. A spotlight platform is visible, just to the left of the Commandant's home, and saluting guns are center in the foreground.

Center House and the row of officers quarters lining the 8th Street side of the quadrangle are shown in this photograph. Center House continues to be the center of social activities for guests of the barracks and also serves as a Commissioned Officers' mess.

This modern-day entrance gate to the "Oldest Post in the Corps" is located appropriately at the corner of 8th and I Streets.

RECOMMENDED READING

Millet, Allan R. *Semper Fidelis: The History of the United States Marine Corps*. New York: Macmillan, 1980.

Moskin, J. Robert. *The U.S. Marine Corps Story*. New York: McGraw-Hill, 1987.

Simmons, Edwin H. *The Marines*. Quantico, VA: Marine Corps Historical Foundation and Levin Associates, 1998.